WATER POLLUTION MONITORING

VOLUME 1, ENVIRONMENTAL MONOGRAPH SERIES

Edited by Frank L. Cross, Jr.

a TECHNOMIC® publication
TECHNOMIC Publishing Co., Inc.
265 W. State St., Westport, Conn. 06880

WATER POLLUTION MONITORING

265 W. State St., Westport, CT 06880

a TECHNOMIC® publication

Printed in U.S.A.

Library of Congress Card No. 74-33841

ISBN-0-87762-180-2

CONTENTS

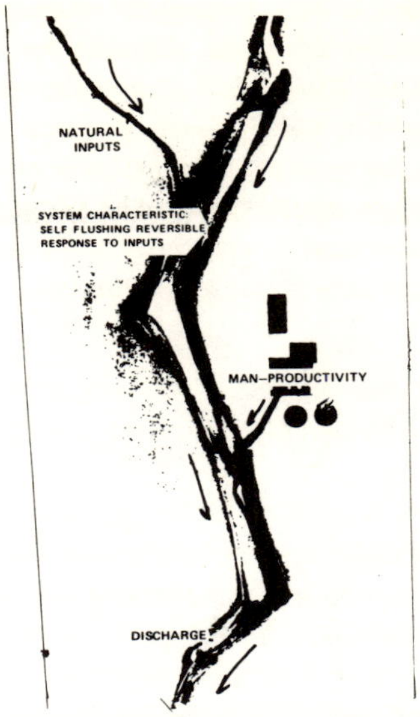

Figure 1. Flowing stream system [1].

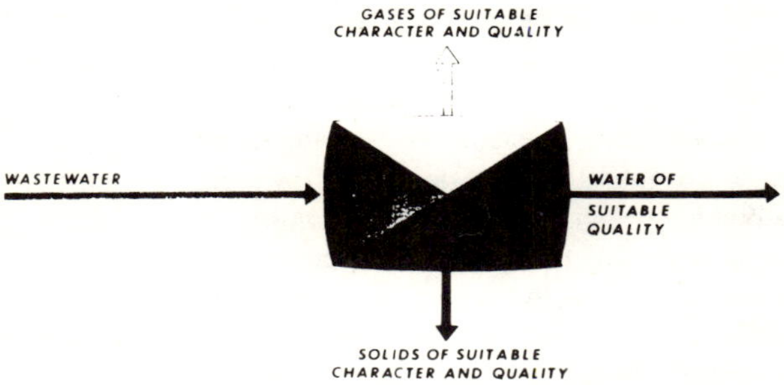

Figure 2. Pollution abatement facility [1].

which is defined as the collection of information on a continual basis, will provide this information.

Monitoring must provide information about the quality and characteristics of wastewater delivered to the system. It must then provide data for man to ascertain that the effluent objectives are being met by the treatment process and that other emissions for the system are suitable in character and quality. Monitoring information must be responsive to operating conditions and must be keyed to critical operating relationships.

For example, in the operation of an activated sludge system, removal efficiency is not the same for all types of materials. Therefore, process performance can effectively be based on the removal of a specific material which is critical to the water quality and found among those materials removed at lower kinetic rates. Also, in the operation of an ionic-exchange system, the objective might be to remove phosphate ion. The logical monitoring would be for phosphate ion, and monovalent ions such as chloride and nitrate might be permitted to pass through the system.

The quality of water in a flowing surface stream might be assured by adequate control of all inputs, but this approach might not be entirely practical or economically feasible. Some of the parameters that might be monitored in a flowing surface water stream are:

1. Low flow augmentation
2. Possible forms of in-stream treatment
3. Wastewater impoundment for controlled releases
4. Control of surface runoff and treatment
5. Variable operation of wastewater treatment facilities, etc.

EFFLUENT AND STREAM MONITORING

INTRODUCTION

The discussion of monitoring in this monograph is related to water systems. It is evident that, pollution abatement must be related to the total system including all of the interfaces (i.e., source, carrier and receptor). Water can be considered independently because it is the natural repository of most waste materials.

The physical size of large bodies of water (the seas, oceans, and larger lakes) have led man to the misconception that his activities were too small to have a significant impact on such bodies of water. The magnitude of the misconception is now evident. The Great Lakes in the United States show the impact of man's interface. Likewise, man's impact on the oceans can be detected by an evident gradation of lead through the depth of the ocean waters, where the source of the lead presumably is fumes from internal combustion engines.

Large bodies of water have a large degree of assimilation, but at the same time, the avenues for elimination are significantly limited. Therefore, such bodies of water integrate the accumulation of materials over a long period of time. The response to this accumulation may develop at a relatively slow rate. The route of elimination may progress slowly through deposition of insoluble materials and through adsorption, assimilation, and precipitation of certain soluble materials. Thus, large bodies of water can involve slow response, and thus, in the short-term view have a definite characteristic of irreversibility.

Responses which are slow and reflect long-term accumulations make it necessary to assure a safety margin in control and management of man's impact on large bodies of water.

The impact of man's activities on surface stream water quality has been evident and has received a significant level of attention.

Flowing surface streams are dynamic systems which have rapidly responded to pollution inputs or abatement procedures.

The flowing stream receives natural inputs through direct contact with air, with land which constitutes the channel, and with runoff from natural areas. Additionally, the stream receives contaminated runoff from man's agricultural activities and from the surfaces of urban areas. These inputs are relatively uncontrolled in that they are delivered by the flow established during natural rainfall.

Other man-produced inputs include those sources of industrial wastewaters and municipal wastewaters that are collected and delivered to the receiving stream. The condition of water quality in our surface streams is largely a man-produced problem. The correction of that problem will definitely entail the control of land pollution, air pollution and wastewater treatment.

Man's efforts must significantly be related to the elimination of pollution problems at the source, and monitoring will play a significant role in that effort.

Wastewater treatment processes either remove and concentrate contaminants, or change the contaminant to innocuous materials. A wastewater treatment facility must interface successfully with the total environment. To do so, any gases emitted must be of acceptable quality and character, and the same must be true of any solids discharged.

Monitoring, Surveillance and System Management are necessary for man to have adequate information to understand the impact of pollutants upon water systems and to develop corrective measures. Monitoring,

Quality of water in the flowing stream is very frequently related to the discharge volume in the stream. During dry weather periods, the low volume of water available for dilution can result in undesirably high concentrations of certain pollution materials. Also, thermal pollution problems might relate directly to volume and temperature of water available. It might be economically practical to consider the installation of impoundments or of other facilities and/or techniques to provide low-flow augmentation.

There appears to be little potential for manipulative control strategies in large bodies of water. Therefore, the objective must be to monitor the water quality and provide the level of surveillance which can insure that objectives are being obtained so that proper objectives can be reached.

Thus far we have seen that:

1. Monitoring is basic to successful water quality control and management.
2. Operation of wastewater treatment facilities should utilize monitoring inputs as to quality, quantity and kinetics.
3. Time constants in treatment processes are normally short, therefore, monitoring information is required essentially instantly and definitely with delays of no longer than a few minutes.
4. Process control should be related to critical specific components rather than (or in combination with) gross parameters. Therefore, monitoring for specific organics or inorganic materials is indicated.
5. Flowing surface streams are potentially manageable, based on continual monitoring information. The system is responsive, provided that monitoring information is available within minutes, or at least within a few hours.
6. Large bodies of water are not readily manipulated to adjust the quality of water. However, monitoring is required for surveillance purposes.

STREAM AND WASTE GAUGING [2]

The initial step in developing a sound water quality control program is the metering, sampling, and analysis of all the significant streams leading into a plant, circulating through the plant, and leaving the plant. This procedure is basic to any of the following purposes:

1. To meet regulatory agency effluent requirements
2. To determine equitable charges for joint treatment
3. To design adequate treatment processes
4. To plan for water reuse and conservation
5. To determine produce losses, either for the purpose of planning a recovery system or for the design of processes which prevent such losses
6. To determine treatment efficiency and to trouble-shoot any process failures
7. To design treatment plant expansion
8. To serve as the basis for development of an automatic treatment system.

The importance of gauging and sampling in the conduct of a stream or waste survey cannot be over-emphasized. When one pauses to consider the cost of obtaining a sample or discharge measurement, it becomes readily apparent that equipment and techniques should receive careful consideration. If the analysis of survey data involves quantitative considerations, the accurate measurement of discharge assumes a level of importance equal to that of the laboratory and analytical results. (Flow X concentration X factor = pounds.) Similarly, the analytical results are only as representative as the sample upon which the analysis is performed. Success or failure in obtaining a representative sample is dependent upon the adequacy of the equipment, and techniques employed in its collection.

In the following discussion, various sampling and gauging procedures are described which have been found to be applicable in field studies. In addition, an attempt has been made to provide information that will aid in the selection of equipment and methods.

GAUGING

A. Gauging Programs

Gauging stations are normally established at or near sampling stations in order to determine discharge concurrent with the collection of samples. The importance of accurate discharge measurements makes the careful establishment of gauging stations imperative. The time and effort involved in obtaining these measurements is a factor requiring that a minimum number of stations be established consistent with adequate definition of flow patterns.

B. Stream Gauging

1. Current Meter.—The current meter is a device for measuring the velocity of a flowing body of water. The stream cross-section is divided into a number of sections, and the average velocity in each section is determined. The discharge is then found by summing the products of area and velocity of each section.

TABLE 1
Methods of Flow Measurement and Their Application to Various Types of Problems

Device or Method	Flow Range Measurement	Cost	Ease of Installation*	Accuracy of Data	Application
Mathematical formulas	Small to large	Low	N/A	Fair	Open channels and pipe flow.
Water meters	Small to large	Low	Fair	Excellent	Pipe flow.
Bucket and stopwatch	Small	Low	N/A	Good	Small pipes with ends or where joints can be disconnected.
Pump capacity and operation	Small to large	Low	N/A	Good	Lines where water is being pumped.
Floating objects	Small to medium	Low	N/A	Fair	Open channels.
Dyes	Small to medium	Low	N/A	Fair to average	Pipe flow and open channels.
Salt dilution	Small to medium	Low	N/A	Fair	Pipe flow and open channels.
Orifice meter	Small to large	Medium	Fair	Excellent	Pipe flow.
Weirs and flow recorders	Small to large	Medium	Difficult	Good to excellent	Open channels.
Parshall Flume	Small to large	High	Difficult	Excellent	Open channels.
Venturi meter	Small to large	High	Fair	Excellent	Pipe flow.
Magnetic flow meter	Small to large	High	Fair	Excellent	Pipe flow.
Flow nozzles	Small to large	Medium	Fair	Excellent	Pipe flow.
Pitot tube	Small to medium	Medium	Fair	Good	Pipe flow.
Rotameter	Small to medium	Medium	Fair	Excellent	Pipe flow.

*N/A = Not Applicable

If the gauging station is properly chosen and a sufficient number of observations are made, a stage vs. discharge relationship may be determined. This method is recommended when the quantity of water to be measured is large and the available fall is small. It is also useful for measuring smaller discharges when it is impractical or uneconomical to employ such devices as a weir or flume.

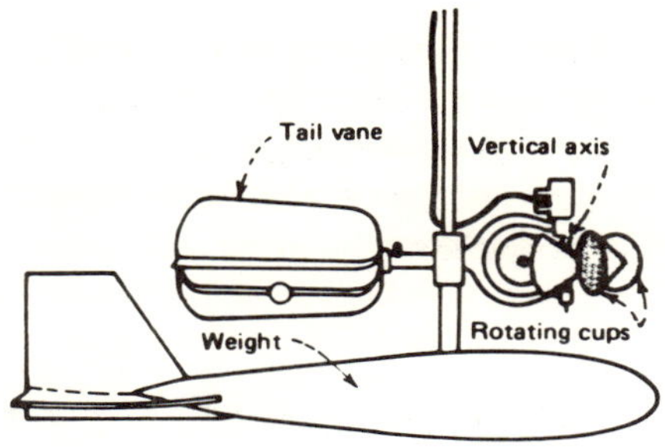

Vertical axis current meter

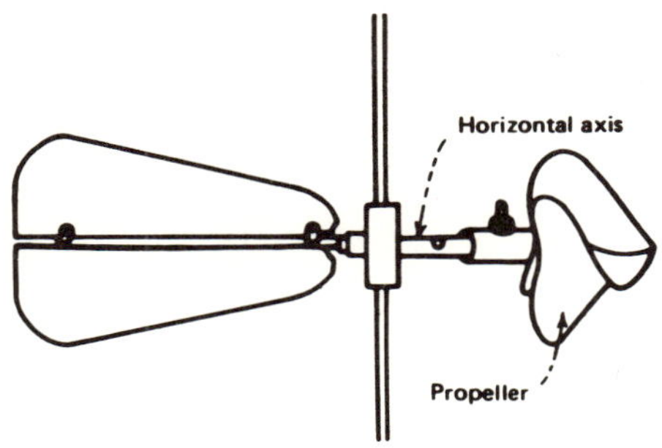

Horizontal axis current meter

Figure 3. Current meters are made either with vertical axis (top) or horizontal axis (propeller type). [2]

2. Weirs.—A weir may be defined as a bulkhead across a channel, in the top of which is a notch through which liquid flows. This notch, of regular shape and dimensions, is called the weir notch, and its bottom edge is known as the crest. The sheet of liquid passing through the notch and falling over the weir crest is known as the nappe. The difference in elevation between the crest and the liquid surface at a point upstream, a specified distance from the bulkhead, is known as the head of the weir. Head discharge relationships have been developed for the several types of weirs.

Weirs are especially adapted to situations where a continuous or semi-continuous discharge record is desired. The principal disadvantages of the weir are the labor required in installation, and the fact that, if the water is high in solids, siltation in the weir pond may destroy its usefulness.

(a) Standard Contracted Rectangular Weir.—This type of weir has its crest and sides so far removed from the bottom and sides of the channel that full contraction of the nappe is developed.

(b) Standard Suppressed Rectangular Weir.—This type of weir has its crest so far removed from the bottom of the channel that full crest contraction is developed. The sides of the weir are coincident with the sides of the channel so that no lateral contraction of the nappe is possible.

(c) V-Notch Weirs.—In this type of weir, the notch is in the shape of a "V" with the sides of the notch being inclined equally from the vertical. The north angle may be of almost any magnitude, but is usually either 60° or 90°. This type of weir is especially suited to the accurate measurement of small discharges since more head is developed than for the same discharge with a rectangular weir. It is to be preferred for discharges less than 1 cfs and is at least as accurate as any other type of weir for discharges up to 10 cfs.

(d) Submerged Weirs.—If it is impossible to use one of the above weirs (because of insufficient available fall, for example), a submerged weir may be employed. This is essentially a standard suppressed rectangular weir in which the water surface downstream is above the crest level.

There are certain conditions of installation and operation which must be satisfied if weir measurements are to be accurate. Most hydraulic handbooks and textbooks provide this information along with discharge formulas and tables for the several types of weirs. Frequent errors in weir installation include failure to provide proper ventilation of the nappe and insufficient attention to horizontal and vertical installation. It is essential that the installation of the weir comply with recommended procedures if results are to be reliable.

3. Parshall Flume.—The Parshall flume is an open constricted channel in which differences in elevation of the liquid flowing through are translated into rate of flow. It consists of an entrance section with converging vertical walls and level floor, a throat section with parallel walls and floor declining downstream, and an exit section with diverging vertical walls and floor inclining downstream. The Parshall flume is especially adapted to situations where a continuous or semi-con-

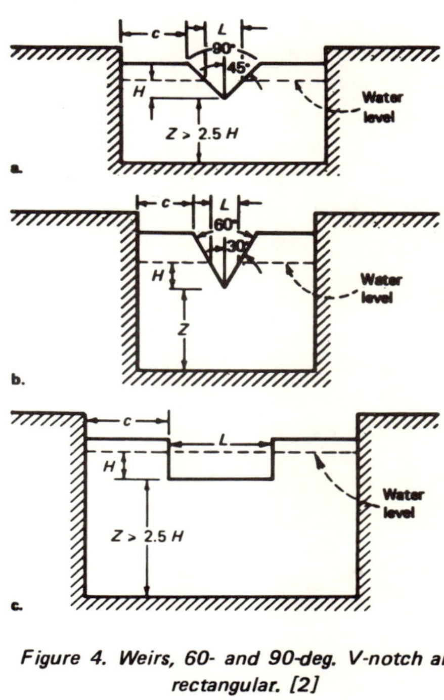

Figure 4. Weirs, 60- and 90-deg. V-notch and rectangular. [2]

tinuous discharge record is desired. Its most significant advantages are low head requirements, very little effect from approach velocity, dependable accuracy, wide variations in capacity, and the fact that, because it is self-cleaning, its usefulness will not be impaired by water high in solids. The primary disadvantage is the cost of installation, which may be significantly higher than for weirs. This high installation cost is occasioned, at least in part, by the precision required if accurate discharge measurements are to be expected.

C. Gauging Conduits and Open Channels

1. Weirs and Parshall Flumes.—Weirs and Parshall flumes are very well adapted to gauging conduits and open channels.

TABLE 2

Measuring Technique	Type of Monitoring	
	Effluent	Stream
Weirs and Parshall flumes	X	X
Bucket and Stopwatch	X	—
Velocity Area Methods		
Current meter	—	X
Floats	X	X
Dye	X	X
Depth Slope	X	—
Open End Pipe Flow	X	—
Head Measuring Devices	X	X

2. Bucket and Stopwatch.—In this method, the time required to fill a container of known volume is measured. This is then translated into a rate of flow. This method should be confined to the measurement of small discharges. A modification of it involves the use of a tank in the process which carries the full flow to be measured. The liquid level in the tank may be dropped a known amount and the time required to refill the tank measured. Knowing the volume of the portion of the tank that was refilled, the rate of flow may be computed. Unless the volume of the container to be filled is such that at least 20 sec are required for filling, the accuracy of this method is very questionable.

3. Velocity-Area Methods.—These methods of discharge measurement involve the determination of the velocity and the cross-sectional area of the flowing liquid. The discharge is then computed as the product of velocity and area.

(a) Current Meter.—The current meter may be employed in the same manner as in stream gauging. However, the presence of foreign materials such as rags, fibers, or slimes may reduce the accuracy of velocity readings. Also, corrosive waters may damage the meter. If the depth of flow is slight, it may be difficult, if not impossible, to use a current meter.

(b) Floats.—Floats may be used to estimate the time of travel between two points a known distance apart. The velocity so obtained may be multiplied by 0.85 to give the average velocity in the vertical. Knowing the mean velocity and the area of the flowing stream, the discharge may be estimated. Floats should be employed only when other methods are impractical.

(c) Dye.—A rough estimate of velocity of flow may be obtained by injecting dye into the flowing stream and timing its travel between two points a known distance apart. Discharge may then be estimated if the area of the flowing stream is known.

(d) Depth-Slope.—If the depth of the flowing stream and the slope of the sewer invert are known, the discharge may be computed by means of any one of several pipe flow formulas.

4. Open End Pipe Flow.—The following methods can be employed when other more precise means are not practical. They can be employed, however, only when there is free discharge to the air.

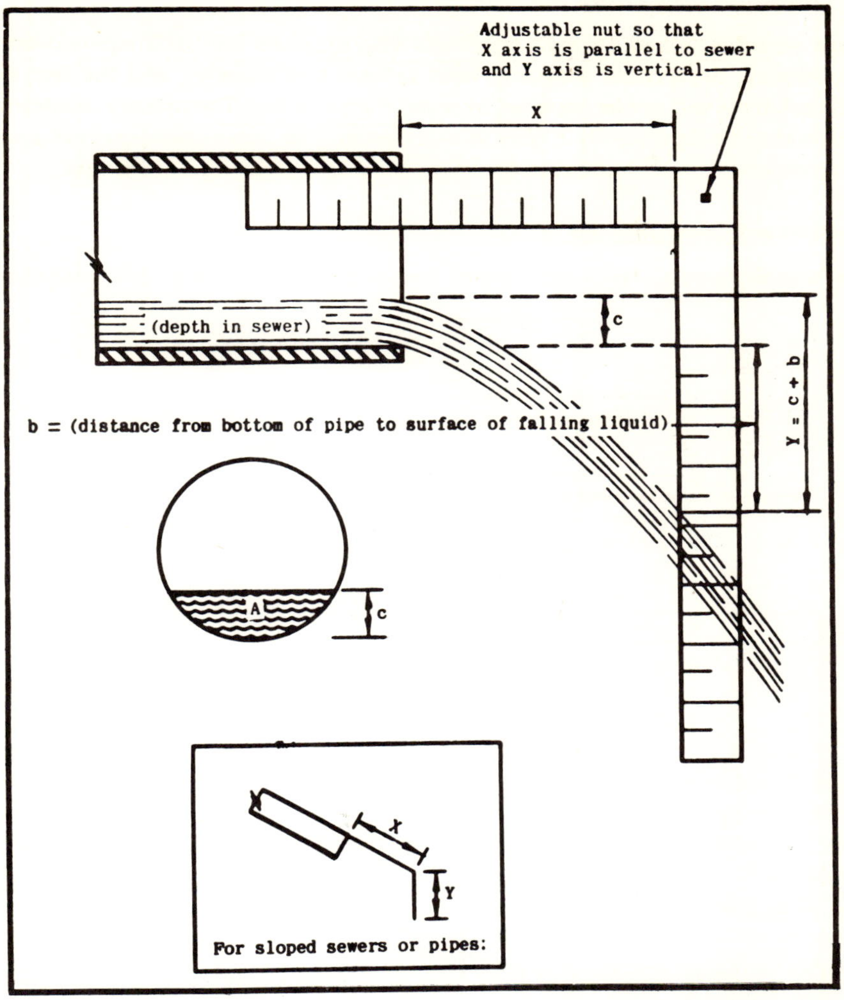

Figure 5. Open-pipe flow measurement — this device, adjusted to the slope of a sewer and calibrated, can then be clamped to the sewer outfall.

(a) Coordinate Method.—Discharge may be computed by the following formula:

$$Q\ (\text{gpm}) = \frac{1800\ AX}{Y}$$

where: A = cross-sectional area of liquid in the pipe (sq ft)
X = distance between the end of the pipe and the vertical gauge in ft, measured parallel to the pipe
Y = vertical distance from water surface at the end of the pipe to the intersection of the water surface with the vertical gauge, in ft

(b) California Pipe Flow Method.—This method may be used only for horizontal pipes having free discharge. If the pipe is not horizontal, a connection must be made to one that is. The horizontal length must be not less than 6 times the diameter of the pipe. Discharge may be computed by the following formula:

$Q\ (\text{gpm}) + T \times W$

where $T = 3{,}900\ (1 - \frac{a}{d})\ 1.88$

$D = d^{2.48}$

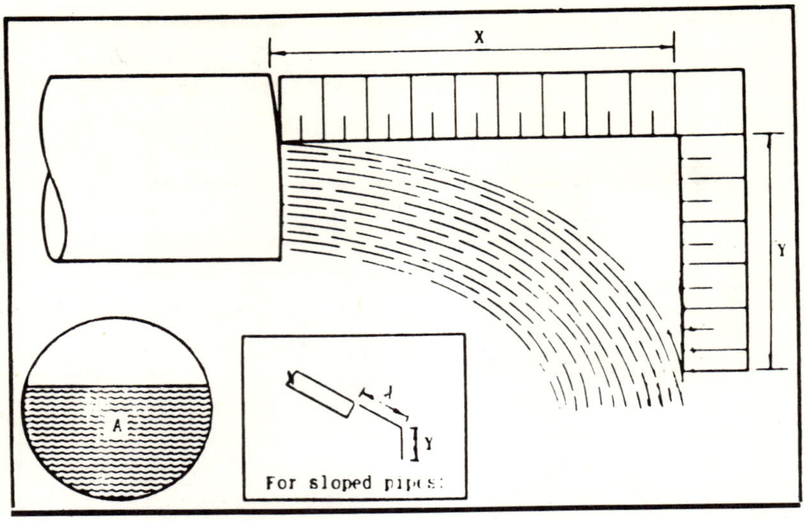

Figure 6. Open-pipe flow measurement requires two dimensions that locate the surface of stream after it leaves the pipe.

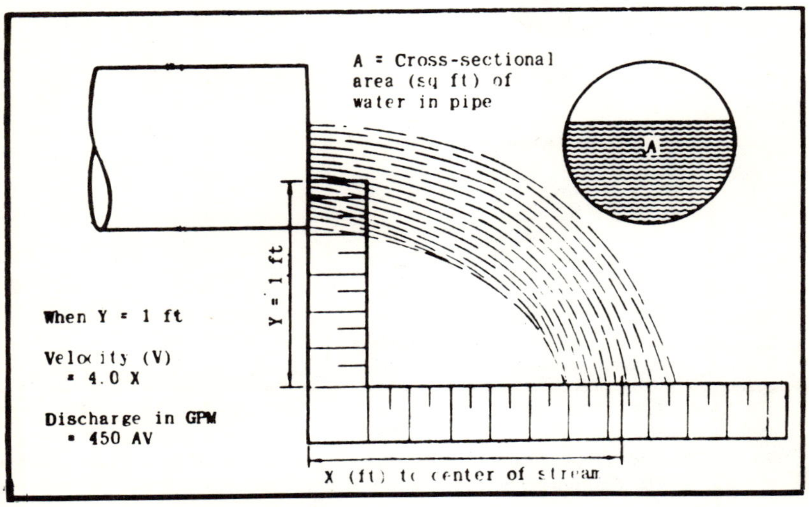

Figure 7. How to measure velocity and discharge from a pipe.

5. Dilution.—Flow may be measured by adding to the flow to be measured a solution containing a known concentration of some chemical. This solution must be added at a constant known rate. The concentration of the chemical at some point downstream is measured after adequate mixing has taken place. If the natural concentration of the chemical in the waste stream is known, the discharge may be computed. Salt is most commonly used for this purpose, but any chemical easily identified and not subject to reaction in the waste stream may be used. The rate of addition of the solution must be negligible compared to the discharge to be measured. Assuming that the rate of addition has no effect on the total discharge, the computation to be made is as follows:

$$Q_1 = \frac{Q_2 C_2}{C_2 C_1}$$

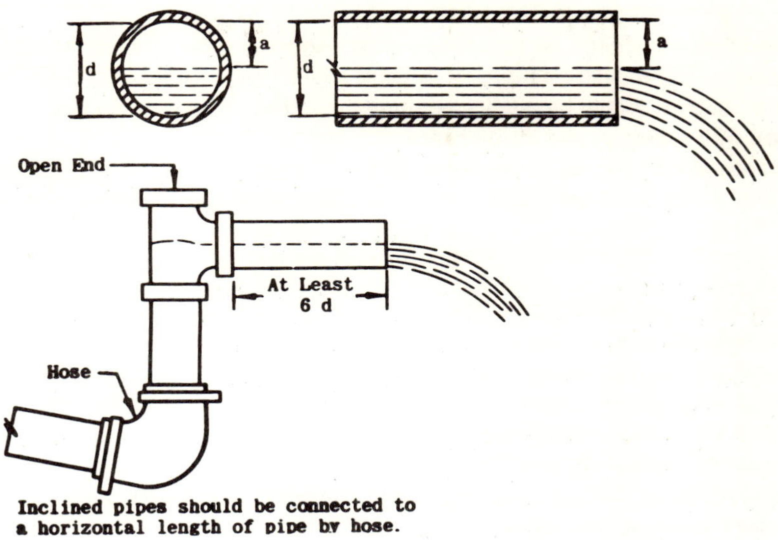

Figure 8. Measurements needed for California Pipe Flow Method.

where: Q_1 = Discharge of waste stream
Q_2 = Rate of addition of solution
C_1 = Concentration in waste stream above point of addition
C_2 = Concentration in solution added
C_3 = Concentration in waste stream below point of addition

D. Head Measuring Devices

Several of the above gauging methods require the measurement of water level in order that discharge may be determined. Any device used for this purpose must be referenced to some zero elevation. For example, the zero elevation for weir measurements is the elevation of the weir crest. The choice of method is dependent upon the degree of accuracy and the type of record desired.

1. Hook Gauge.—The hook gauge measures water elevation from a fixed point. The hook is dropped below the water surface and then raised until the point of the hook just breaks the surface. This method probably will give the most precise results when properly applied.

2. Staff Gauge.—The staff gauge is merely a graduated scale placed in the water so that elevation may be read directly.

3. Plumb Line.—This method involves measurement of the distance from a fixed reference point to the water surface by dropping a plumb line until it just touches the water surface.

4. Water Level Recorder.—This instrument is used when a continuous record of water level is desired. A float and counterweight are connected by a steel tape which passes over a pulley. The float should be placed in a stilling well. A change in water level causes the pulley to rotate which, through a gearing system, moves a pen. The pen traces water level on a chart which is attached to a drum that is rotated by a clock mechanism. When properly installed and maintained, the water level recorder will provide an accurate, continuous record.

E. General Considerations

This discussion has been limited to those methods of flow measurement that have been found to be most consistent with the requirements of field studies. Measurement of flow in pipes under pressure has been omitted since measurements of this type are seldom required in the course of a short term field study. Similarly, many acceptable methods of measuring free surface flow have been omitted in the belief that the methods discussed herein should be adequate a great majority of the time. It should be emphasized, however, that the ingenuity of the engineer is his most valuable tool in the solution of unusual problems.

The choice of method should be predicated upon the following considerations:

1. Ease and cost of installation and maintenance
2. Obtainable accuracy in relation to desired accuracy
3. Adequacy (e.g., water level recorder should be used if a continuous record is desired).

SAMPLING METHODS

The basis for any plant pollution abatement program rests upon information obtained by sampling. Thus, all subsequent decisions may be based upon incorrect information if this step is not accurately pursued. There are several pitfalls which can occur if sampling is performed in a careless or naive manner. If a few basic principles are followed and if those responsible for sampling are forewarned, reliable results can be obtained without extensive and costly resampling.

Obtaining good results will depend upon certain details. Among these are the following:

1. Insuring that the sample taken is truly representative of the wastestream,
2. Using proper sampling techniques,
3. Protecting the samples until they are analyzed.

Obviously, the type of sample and sampling equipment will depend upon:

A. Pollutants to be monitored
B. Quantity of sample required
C. Frequency of sampling
D. Character of liquid (fluid) being sampled
E. Nature of the sampling station

TABLE 3

Volume of Sample Required for Determination of the Various Constituents of Industrial Water [2]

	Volume of Sample,[a] ml
PHYSICAL TESTS	
*Color and Odor	100 to 500
*Corrosivity	flowing sample
*Electrical conductivity	100
*pH, electrometric	100
Radioactivity	100 to 1000
*Specific gravity	100
*Temperature	flowing sample
*Toxicity	1000 to 20 000
*Turbidity	100 to 1000
CHEMICAL TESTS	
Dissolved Gases:	
†Ammonia, NH_3	500
†Carbon dioxide, free CO_2	200
†Chlorine, free Cl_2	200
†Hydrogen, H_2	1000
†Hydrogen sulfide, H_2S	500
†Oxygen, O_2	500 to 1000
†Sulfur dioxide, free SO_2	100
Miscellaneous:	
Acidity and alkalinity	100
Bacteria, iron	500
Bacteria, sulfate-reducing	100
Biochemical oxygen demand	100 to 500
Carbon dioxide, total CO_2 (including CO_3^{-}, HCO_3^{-}, and free)	200
Chemical oxygen demand (dichromate)	50 to 100
Chlorine requirement	2000 to 4000
Chlorine, total residual Cl_2 (including OCl^-, HOCl, NH_2Cl, $NHCl_2$, and free)	200
Chloroform - extractable matter	1000
Detergents	100 to 200
Hardness	50 to 100
Hydrazine	50 to 100
Microorganisms	100 to 200
Volatile and filming amines	500 to 1000
Oily matter	3000 to 5000
Organic nitrogen	500 to 1000
Phenolic compounds	800 to 4000

(continued)

TABLE 3
(Continued)

	Volume of Sample,[a] ml		Volume of Sample,[a] ml
Miscellaneous:		*Cotions:*	
pH, colorimetric	10 to 20	Nickel, Ni^{++}	100 to 1000
Polyphosphates	100 to 200	Silver, Ag^{+}	100 to 1000
Silica	50 to 1000	Sodium, Na^{+}	100 to 1000
Solids, dissolved	100 to 20 000	Strontium, Sr^{++}	100 to 1000
Solids, suspended	50 to 1000	Tin, Sn^{++} and Sn^{++++}	100 to 1000
Tannin and lignin	100 to 200	Zinc, Zn^{++}	100 to 1000
Cotions:		*Anions:*	
Aluminum, Al^{+++}	100 to 1000	Bicarbonate, HCO_3	100 to 200
†Ammonium, NH_4^{+}	500	Bromide, Br^-	100
Antimony, Sb^{+++} to Sb^{+++++}	100 to 1000	Carbonate, CO_3^{--}	100 to 200
Arsenic, As^{+++} to As^{+++++}	100 to 1000	Chloride, Cl^-	25 to 100
Barium, Ba^{++}	100 to 1000	Cyanide, Cn^-	25 to 100
Cadmium, Cd^{++}	100 to 1000	Fluoride, Fl^-	200
Calcium, Ca^{++}	100 to 1000	Hydroxide, OH^-	50 to 100
Chromium, Cr^{+++} to Cr^{++++++}	100 to 1000	Iodide, I^-	100
Copper, Cu^{++}	200 to 4000	Nitrate, NO_3^-	10 to 100
†Iron, Fe^{++} and Fe^{+++}	100 to 1000	Nitrite, NO_2^-	50 to 100
Lead, Pb^{++}	100 to 4000	Phosphate, ortho, PO_4^{---}, HPO_4^{--}, $H_2PO_4^-$	50 to 100
Magnesium, Mg^{++}	100 to 1000	Sulfate, SO_4^{--}, HSO_4^-	100 to 1000
Manganese, Mn^{++} to $Mn^{+++++++}$	100 to 1000	Sulfide, S^{--}, HS^-	100 to 500
Mercury, Hg^{+} and Hg^{++}	100 to 1000	Sulfite, SO_3^{--}, HSO_3^-	50 to 100
Potassium, K^{+}	100 to 1000		

[a] Volumes specified in this table should be considered as a guide for the approximate quantity of sample necessary for the particular analysis. The exact quantity used should be consistent with the volume prescribed in the standard method of analysis, whenever the volume is specified.

* Aliquot may be used for other determinations.

† Samples for unstable constituents must be obtained in separate containers, preserved as prescribed, completely filled and sealed against all exposure.

There are six major types of sampling techniques: individual or grab sample, simple composite, sequential composite, continuous sample, hand proportioned composite, and automatically proportioned composite. *The individual or grab sample* is retained as a separate entity in its own container. In the extreme case, for very small flows, the entire stream can be collected during the time interval selected, then mixed, and an aliquot taken for analysis.

The simple composite requires that all samples taken over the specified time interval be deposited in a single container. If, however, the composite is to be analysed for oily materials, it must be divided between two separate containers. This is true also if different methods of preservation are required (other than refrigeration) for the proposed analyses.

The sequential composite requires the collection of a series (usually 2 to 8) of individual samples per container, each container representing a specific time period, generally one hour. Such a procedure is particularly useful where the character of the waste may vary significantly from hour to hour, where batch dumping is expected, or where self-cancelling conditions occur, such as alternating high and low pH, which would not be apparent in a simple composite sample.

The continuous sample in which a small amount of sample is collected in continuous flow is useful for feeding monitors or pilot scale processes, or for sampling receiving waters. This technique is not recommended where the stream is high in suspended solids unless special provisions are made to avoid settling and plugging or partial plugging of the lines. If the continuous sample is composited over a long period of time, the product is large, at times too large to be conveniently handled.

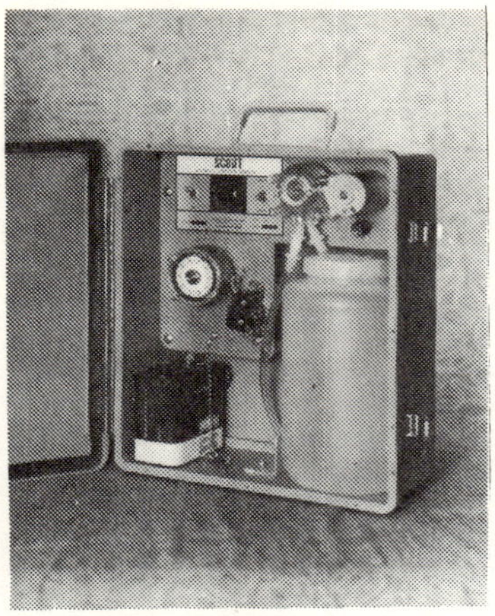

Figure 9. Simple composite sampling unit.
Source: N-CON Systems Co., Inc.
New Rochelle, NY

Figure 10. Sequential composite sampling unit.
Source: N-CON Systems Co., Inc. — New Rochelle, NY

Figure 11. Pump supplies a continuous sample to sensors.
Source: Honeywell Industrial Division
1100 Virginia Drive
Fort Washington, Pennsylvania 19034

The hand proportioned composite can be obtained where flow charts are available. Individual or sequential composite samples are manually composited in proportion to the flow to obtain this useful and representative type of sample.

The automatically proportioned composite does require additional equipment and an understanding of the principles of proportioning, but is not especially complicated. Sampling in proportion to flow may be accomplished in either of two ways: By varying the frequency of sampling in direct ratio to flow (VcTv, volume constant, time variable), or by varying the size of the individual aliquots in proportion to flow volume (VvTc, volume variable, time constant).

Samples should be analyzed as soon as possible after collection. However, in practice, immediate analysis is seldom feasible. Cognizance should be taken of certain time-dependent chemical changes that can occur in samples, such as:

1. Metal cations may precipitate as hydroxides or form complexes.
2. The valence state of the ions may change by oxidation or reduction.
3. Metal cations may be adsorbed on the surfaces of glass, plastic, or quartz containers.

Microbiological activity may also change the characteristics of the sample as follows:

1. Cell lysis may increase the BOD and COD.
2. Cell productivity may change the BOD and COD.
3. The organic nitrogen and organic phosphorus content may be changed.

TABLE 4

Suggested Sampling or Compositing Schedule [3]

Characteristic	High Variability	Low Variability
BOD[a]	4 hr	12 hr
COD or TOC[a]	2 hr	8 hr
Suspended Solids	8 hr	24 hr
Alkalinity or Acidity	1 hr grab	8 hr grab
pH	Continuous	4 hr grab
Nitrogen and Phosphorus[b]	24 hr	24 hr
Heavy Metals	4 hr	24 hr

[a]The compositing schedule where continuous samplers are not used depends on variability, i.e., 15 min for high variability to 1 hr for low variability.

[b]Does not apply to nitrogen or phosphorus wastes (e.g., fertilizer).

In a water quality management program, automatic monitoring of several water characteristics has proven to be a dependable method of control. In wastewater treatment, numerous parameters are used for operational control; however, the number of parameters that can be automatically measured without difficulty are limited.

Sensors for automatic monitoring of wastewaters are especially sensitive to the presence of interferences. Thus, great care should be exercised in the selection of automatic equipment in order to ensure that it will function satisfactorily in the wastewater to be monitored.

Automatic monitoring has the following advantages [3, 4]:

1. The parameters of interest are recorded on a continuous basis and a clear picture is obtained of the variation of the recorded parameters with time. It should be noted, however, that continuous flow measurement data must also be available in order to calculate the total amount of pollutants flowing on a daily basis.
2. There is a shorter time lag between sampling and analysis than in manual sampling. In addition, problems resulting from storage of samples are eliminated.
3. Automatic monitoring systems can be combined with an alarm system that will give advance warning when a high concentration of an undesirable parameter occurs. For example, an automatic conductivity measurement instrument could be set to detect high values. When this occurs a by-pass valve

TABLE 5

Sample Preservation [3]

Parameter	Preservative	Maximum Holding Period
Acidity-Alkalinity	Refrigeration at 4° C	24 hours
Biochemical Oxygen Demand	Refrigeration at 4° C	6 hours
Calcium	None required	7 days
Chemical Oxygen Demand	2 ml H_2SO_4 per liter	7 days
Chloride	None required	7 days
Color	Refrigeration at 4° C	24 hours
Cyanide	NaOH to pH 10	24 hours
Dissolved Oxygen	Determine on site	No holding
Fluoride	None required	7 days
Hardness	None required	7 days
Metals, Total	5 ml HNO_3 per liter	6 months
Metals, Dissolved	Filtrate: 3 ml 1:1 HNO_3 per liter	6 months
Nitrogen, Ammonia	40 mg $HgCl_2$* per liter - 4° C	7 days
Nitrogen, Kjedahl	40 mg $HgCl_2$* per liter - 4° C	Unstable
Nitrogen, Nitrate-Nitrite	40 mg $HgCl_2$* per liter - 4° C	7 days
Oil and Grease	2 ml H_2SO_4 per liter - 4° C	24 days
Organic Carbon	2 ml H_2SO_4per liter (pH 2)	7 days
pH	Determine on-site	No holding
Phenolics	1.0 g $CuSO_4$/1 + H_3PO_4 to pH 4.0 - 4° C	24 hours
Phosphorous	40 mg $HgCl_2$* per liter - 4° C	7 days
Solids	None available	7 days
Specific Conductance	None required	7 days
Sulfate	Refrigeration at 4° C	7 days
Sulfide	2 ml Zn acetate per liter	7 days
Threshold Odor	Refrigeration at 4° C	7 days
Turbidity	None Available	7 days

could be opened and the waste stream directed to a storage basin from which it could be gradually added into the waste treatment system.

4. Instrumentation, professionally installed, properly maintained to insure good operating order provides extreme precision of operation and optimum utilization of manpower and facilities. For example, the time lag encountered for an operator to get to a remote station to make a change is eliminated . . . all system adjustments can be effected immediately when and where needed from a centralized location.
5. The value of indication and the recording of key system data cannot be discounted lightly. When this information is received at a central operating point in the facility, the operator knows exactly what is happening at that time at each of various control points in the system. Thus, not only can operating conditions be effectively analyzed but also this information can be invaluable for planning future operations.
6. A complex, multi-unit system can be operated with fewer personnel. Continuous, reliable measurements of key variables such as flow, level, and pressures in pump control and distribution systems reduce pumps, piping and storage to the level of present need.
7. Reduced mileage in automotive service units necessary for maintenance crews over a year's time period can accomplish material savings in operating costs. Operation can be maintained and continued at times when remote stations may be inaccessible due to flooding, inclement weather, etc.
8. Continuous water quality monitoring and control of specific parameters (hardness, turbidity, chlorine residual) results in higher efficiency by increasing thru-put rates at higher quality levels. Modern instrumentation has made a significant contribution towards permitting design and treatment operations to accommodate loading rates at more than double the traditionally established values of the past.

Disadvantages of automatic monitoring, are:

1. The sensor of the system may not be capable of registering unusual circumstances that occur at the place of sampling.
2. The initial cost of automatic monitoring is high.
3. The wastewater characteristics need to be known before installing automatic monitoring equipment.
4. At present, only relatively simple continuous measurements are dependable, such as pH, temperature, conductivity, and dissolved oxygen.

Automatic monitoring can be of great value when it is combined with the operation of treatment facilities. The sensor could be a simple electrode, such as is used to monitor pH, or it may be a much more sophisticated piece of equipment. Normally, data is collected on a strip chart recorder; however, other appurtenances, such as pumps or valves, may also be activated by the sensor. Problems to be anticipated when using automatic monitoring equipment include:

1. Loss of calibration. Regular maintenance is necessary to prevent errors.
2. The flow system and sensor may fail to operate correctly when suspended bacteria are permitted to grow. Therefore, regular cleaning of the system is necessary. Self-cleaning sensors are available and have a definite advantage when used in situations requiring frequent cleaning.
3. Mechanical damage may occur if the intake system or the sensor is not protected by a screen.
4. Miscellaneous problems can be expected to result from power failure, mishandling, pump difficulties, vandalism, etc.
5. Interferences cause many problems and should be known before installing the monitoring system.

BASIC SYSTEM(S) INVOLVED IN WATER POLLUTION MONITORING

Most water pollution monitoring systems, whether simple or complex in nature, include one or more of the following elements:

- sensing element or collection device
- analysis of sample
- read-out of display of data (on-site)
- Telemetry
- read-out or display of data (remote)
- analysis of data

Water quality monitoring is usually conducted by one of the following techniques:

- on-site automatic analyzers
- portable multi-parameter analyzers
- portable single test instruments and chemical test kits
- mobil laboratories
- grab sample in the field followed by laboratory analysis

Most automatic stations include a pump system and sensors that require periodic maintenance. A realistic schedule for automated equipment is illustrated in Table 6.

In order to determine the equipment to purchase for a monitoring station, or the field tests to perform during a stream survey depends upon the purpose of the monitoring and the scope of work.

The purpose of monitoring may be many fold, but may include:

- Comparison of discharge with effluent standards.
- Existing conditions in a reach of a stream.
- Determining the removal efficiency of a control device/system.
- Estimation of the environmental impact of a discharge on a stream.
- Automation of plant operations.

TABLE 6

Monitoring Station — Operation/Maintenance Schedule

Piece of Equipment	Maintenance Period (Months)	Comments
Pumps	3	Pump efficiency can be improved by proper lubrication, limiting use to non-corrosive waters, and providing adequate screening to prevent erosion of the impeller blades. Maintenance includes bearing and impeller checks/replacement.
DO Meters	3	Periodic electronics checks on calibration are required. Probe membrane replacement is often necessary and occasional replacement of seals.
Sensors		As with most "sensors" the following may be required:
Temperature	6-12	o Changing recording charts and refilling pens.
Conductivity	6-12	o Probe cleaning/replacement
pH	4- 5	o Calibration checks
		o Corrosion inspection.

Some of the parameters used in the evaluation of water quality, are listed in Table 7.

A major item in any monitoring system will be the costs for analytical measurements. There are several ways these charges may be minimized, yet meet all regulatory and in-plant requirements, the most important being proper selection of parameters to be monitored. Parameters which have significant pollution potential for selected industries are presented in Table 8.

The requirements for monitoring outfalls will be specified by the regulatory agency. Some of the necessary analyses will be time-consuming and relatively expensive. In many cases, the regulatory agencies will approve a substitution of a less expensive analytical technique if the parameter requiring the more expensive analysis can be accurately inferred from the simpler analysis. An example of this would be the substitution of COD (Chemical Oxygen Demand) analyses for a portion of the BOD_5 (5 day Biochemical Oxygen Demand) analyses if it can be shown that a satisfactory correlation exists between the two parameters. This will require discussion with the appropriate regulatory agency, but should not be overlooked.

For other than outfall monitoring, the selection of parameters is subject only to the requirement that the control be such that the effluent quality at the outfall is within the permit specifications. It is here that a strong effort should be made to find inexpensive analyses which can provide quick, accurate and correct information of those parameters requiring more expensive analytical techniques. Promptness of analysis is quite important since having the results for early action will greatly simplify control requirements. As an example, conductivity can sometimes be used as an indication of total dissolved solids. This is a simple measurement, and one which gives immediate results. It is absolutely necessary, however, to obtain a correlation between the rapid technique and the standard technique for a specific waste.

SOLUBLE ORGANICS

Soluble, degradable organics cause utilization or depletion of dissolved oxygen by the activity of aerobic bacteria. Most industrial wastewaters contain some soluble organics. Examples are the waste liquors from pulp mills, canning plant wash effluents, meat packing wastes, textile scouring and dyeing effluents, milk product wastes and fermentation wastes. The quantity of soluble organics can be measured as BOD, COD, TOC (Total Organic Carbon), and TOD (Total Oxygen Demand).

Soluble Constituents that Produce Tastes and Odors

Tastes and odors may be associated with: 1) decaying organic matter; 2) living algae and other microscopic organisms containing essential oils and other odorous compounds; 3) iron and manganese and other metallic products of corrosion; 4) specific organic chemicals, such as phenols and mercaptans; 5) chlorine and its substitution compounds; and 6) biologically nondegradable synthetic organics.

CHART 1
Continuous Monitoring System

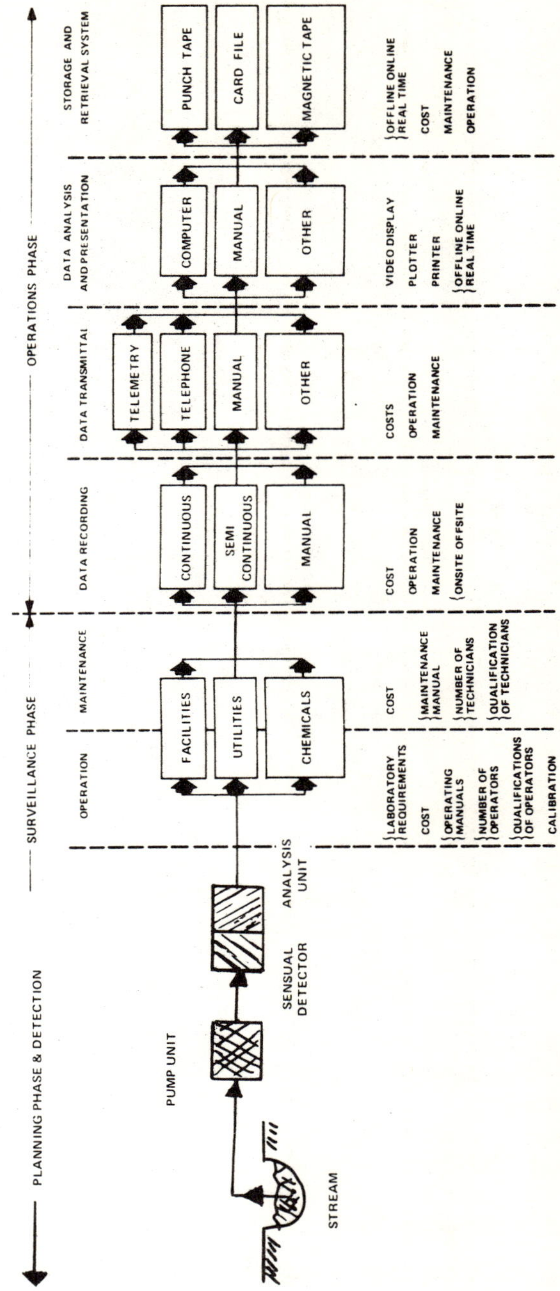

TABLE 7
Parameters Used in Evaluation of Water Quality

Hydrological & Meteorological	Physical	Chemical, Inorganic		Chemical, Organic	Nutrients	Microbiological	Biological
Volumetric Flow Rate	Temperature	Dissolved Oxygen	Trace Elements:	BOD (Immediate,	Organic-N	Coliform, Total &	Plankton
Velocity	Specific Conductance	Dissolved Carbon	Aluminum	(5-day, Long-term)	NH_3-N	Fecal	Periphyton
Time of Flow	Turbidity	Dioxide	Arsenic	COD	NO_2-N	Fecal Streptococci	Benthos
Depth	Light Penetration	Hydrogen Sulfide	Barium	Chlorine Demand	NO_3-N	Total Plate Count	Fish
Tide Variation	Color	Minerals:	Beryllium	Total Organic Carbon	Total-P	Salmonella	Waterfowl
Wind Speed & Direction	Odor	Acidity	Boron	MBAS	Soluble-P	Shigella	Wildlife
Solar Radiation	pH	Alkalinity	Cadmium	CCE	Organic-P	Viruses:	Vascular Plant
Intensity	Total Solids (Volatile	Calcium	Chromium	CAE	Orthophosphate	Coxsackie A&B	
Air Temperature &	& Fixed)	Magnesium	Copper	Cyanide	Polyphosphates	Polio	
Humidity	Suspended Solids	Bicarbonate	Fluoride	Pesticides		Adenovirus	
	(Volatile & Fixed)	Carbonate	Iron	Oil & Grease		Echo	
	Floating Solids	Hydroxide	Lead	Phenolics			
	Sediment	Hardness	Manganese				
	Concentration	Chlorides	Potassium				
	Particle Size	Sulfates	Selenium				
	Bed Load	Dissolved Solids	Silver				
		Radiochemical	Sodium				
			Zinc				

Source: Water Quality Office, EPA.

TABLE 8

Undesirable Characteristics of Industrial Wastewaters [3]

1. Soluble organics causing dissolved oxygen depletion in streams and estuaries.
2. Soluble constituents that result in tastes and odors in water supplies.
3. Toxic materials and heavy metal ions.
4. Color and turbidity.
5. Nutrients—Nitrogen, Phosphorus and Carbon.
6. Refractory materials.
7. Oil, grease and immiscible liquids.
8. Acids and alkalies.
9. Substances resulting in atmospheric odors.
10. Suspended solids resulting in sludge deposits in streams.
11. Dissolved solids.
12. Temperature causing thermal pollution.
13. Radioactive material.
14. Pathogenic wastes.

Phenolics are a special nuisance in drinking water supply, particularly after chlorination, because of the very low concentrations (<2 ppb), which result in taste and odor detection. Petrochemical discharges and liquid wastes from the manufacture of synthetic rubber often cause taste and odor problems, e.g., sulfides cause odors in concentrations less than a few hundredths of a mg/l.

For biological waste treatment plants, the maximum tolerable concentrations of toxic materials have been reported for many materials. Occasionally, treatability studies have to be made to determine the maximum allowable concentration of the toxic substance in a biological treatment system. In general, the threshold toxicity levels for biological treatment systems are higher than the allowable standards for surface waters. Establishing maximum concentrations for toxicants in biological treatment plants is useful only if the amount of toxicant is reduced during the treatment, as is the case with phenols. Often, it is necessary to decrease the concentration of the toxic material by pre-treatment. However, it is necessary to guard against the so-called synergistic effect of certain materials. One plant may be allowed to discharge zinc below the toxic level, while another plant may be allowed to do the same with copper. The resulting combination of both discharges will have a synergistic effect, and may cause biological deterioration in the receiving stream or the municipal treatment system.

Ammonia nitrogen is present in many natural waters in relatively low concentrations while industrial streams often contain exceedingly high concentrations of ammonia. Nitrogen in excess of 1600 mg/l has proven to be inhibitory to many microorganisms present in the activated sludge basin. Sulfides are present in many wastewaters either as a mixture of HS^-—H_2S (depending on pH), sulfonated organic compounds, or metallic sulfides.

The influence of heavy metals on biological unit processes has been the subject of many investigations. Toxic thresholds for Cu, Zn, and Cd, have been established at approximately 1 mg/l.

Color and Turbidity

Color and turbidity present aesthetic problems. Low concentrations of compounds such as lignins and tannins will impart color to natural waters and may be intensified when combined with other materials. An example of this is iron and tannin which combine to form iron-tannate—a common base of blue-black ink.

It is possible to differentiate between the true and apparent color of a sample. True color is due to matter which is in true solution, while apparent color includes the effects of matter in the suspended and colloidal states as well. Examples of true color constituents are soluble dyes used in industry. Constituents which cause apparent color are usually finely divided metal hydroxide particles.

Nutrients—Nitrogen and Phosphorus

When effluents are discharged into lakes, ponds, and surface streams, the presence of nitrogen and phosphorus is particularly undesirable since it enhances eutrophication and stimulates undesirable algae growth. Industrial wastes containing insufficient nitrogen and phosphorus for biological development in waste treatment systems require addition of these nutrients in forms such as anhydrous ammonia and phosphoric acid.

The chemical form in which nutrients are present may differ and vary with the degree of treatment. Nitrogen can be present as ammonia, nitrate, nitrite and organic nitrogen in the form of proteins, urea and amino-acids. Phosphorus can be present as ortho phosphates (for example $Na_3(PO_3)_6$) or organic phosphorus.

Refractory Materials

Refractory materials are resistant to biodegradation and, thus, may be undesirable for certain water quality requirements. Alkyl benzene sulfonate (ABS) from detergents is substantially nondegradable and frequently causes a persistent foam in waste treatment systems and watercourses.

Oil, Grease and Immiscible Liquids

Oil, grease and immiscible liquids can produce unsightly conditions and in most cases the quantities permitted in wastewater are restricted by regulatory agencies. In sewer systems the presence of oils and immiscible liquids, such as napthene and ether, may cause explosive conditions. Wastes from the meat-packing industry, particularly where fats are involved from the slaughtering of sheep and cattle, have resulted in serious decreases in the capacity of sewers. In treatment plants, wastewater with a high grease content may cause trouble with aerobic biological treatment.

The term, grease, applies to a wide variety of organic substances that may be extracted from aqueous solution or suspension by hexane. Hydrocarbons, esters, oils, fats, waxes, and high molecular weight fatty acids are the major materials dissolved by hexane. These materials have a "greasy feel" and are associated with problems in aerobic waste treatment.

Fats, oils, and waxes are esters. Fats and oils are esters of the trihydroxy alcohol, glycerol; while waxes are esters of long-chain monohydroxy alcohols. The glycerides of fatty acids that are liquid at ordinary temperatures are called oils and those that are solids are called fats. The term, oil, also represents a wide variety of substances ranging from low to high molecular weight hydrocarbons of mineral origin, spanning the range from gasoline through heavy fuel to lubricating oils.

It should be emphasized that oils and greases of vegetable and animal origin are generally biodegradable and, in an emulsified form, can be successfully treated by a biological treatment facility. On the other hand, oils and greases of mineral origin may be relatively resistant to biodegradation and will require removal by methods other than biological treatment. Unfortunately, a satisfactory method of distinguishing between oils and greases of vegetable and animal origin and those of mineral origin is not readily available.

Acids and Alkalies

The pH in a biological system, such as in surface water or treatment plants, is an important factor because a sudden change can cause serious damage. The measurement used to determine the required dosage of neutralizing agent, either $Ca(OH)_2$ or H_2SO_4, is termed acidity or alkalinity, respectively.

The acidity measures the capacity to donate protons. Acidity is attributable to the unionized portions of weakly mineral acids, hydrolizing salts and mineral acids. Mineral acids are probably the most significant group. It is difficult to predict neutralization requirements when diverse forms of mineral acidity are prevalent. Microbial systems may reduce acidity in some instances through biological degradation of organic acids.

Alkalinity, or the ability of wastewater to accept protons, is significant in the same general way as acidity, although the biological degradation process does offer some buffer capacity by furnishing carbon dioxide as a degradation end-product to the system. Alkalinity can be due to the presence of HCO_3^-, CO_3^{--}, or OH^-. It has been estimated that approximately 0.5 lb of alkalinity (as $CaCO_3$) is neutralized for each lb of BOD removed. Excess alkalinity often has to be removed by neutralization.

Suspended Solids

Deposition of solids in quiescent stretches of a stream will impair the normal aquatic life of the stream. Sludge blankets containing organic solids will undergo progressive decomposition resulting in oxygen depletion and the production of noxious gases. Identical problems may result when excessive suspended solids are discharged into municipal sewer systems. Suspended solids are normally determined as an organic and an inorganic fraction. The organic fraction is generally distinguished as that portion which is volatilized or oxidized at 550°C.

Dissolved Solids

Many industrial wastes contain high concentrations of dissolved solids and these discharges into surface waters may limit the subsequent use of the stream as drinking or recreational waters. Conventional treatment of the wastewater has little effect on the dissolved solids content. Biological treatment systems are vulnerable to concentrations of salt above 10,000 mg/l.

The concentration ratio of potassium plus sodium and calcium plus magnesium becomes of particular interest if the wastewater is to be used for spray rinsing. The tolerable amounts of sulfates in sewers is also limited because of their corrosive action on concrete and metal sewers as indicated by the following biochemical reaction:

$$SO_4 \xrightarrow{\text{anaerobic}} H_2S \xrightarrow{\text{aerobic}} H_2SO_4$$

Temperature

The addition of heat to surface waters may have deleterious effects. A temperature increase may cause a decrease in the waste assimilative ability of the surface waters. Discharge of heated water to a biological waste treatment plant may have an advantage since bacteria activity increases at higher temperatures with the optimum temperature usually between 30° and 37°C.

Radioactive Material

Radioactive materials can enter a sewerage system or surface waters due to the activities of nuclear reactors and by uranium ore mining and refining. Regulatory agencies have established standards for the maximum allowable concentrations of radioactive materials in surface waters.

It is possible to differentiate between the following three types of radioactivity: α, β, and γ rays. Alpha rays consist of a stream of particles of matter (doubly charged ions of helium with a mass of 4) projected at high speed from radioactive matter. Once emitted in air at room temperature, alpha particles do not travel much more than 10 cm. These particles are stopped by an ordinary sheet of paper.

Beta rays consist of a stream of electrons moving at speeds ranging from 30 to 90 percent of the speed of light, their power of penetration varying with their speed. These particles normally travel several hundred feet in air and may be stopped with aluminum sheeting a few millimeters thick. Gamma rays are true electromagnetic radiations which travel with the speed of light, and are similar to x-rays but have shorter wave lengths and greater penetrating power. Proper shielding from gamma rays requires several centimeters of lead or several feet of concrete. The unit of gamma radiation is the photon.

Radioactive materials commonly used in tracer studies in research in biology, chemistry and medicine are the isotopes of carbon (C^{14}) and iodine (I^{25}). In sewers and waste treatment plants certain isotopes, such as radioiodine and radiophosphorus, acumulate in biological slimes and sludges.

Pathogenic Wastes

Wastewaters that contain pathogenic bacteria can originate from livestock production (cattle, poultry, swine, lab animals), tanneries, pharmaceutical manufacturers and food processing industries. Pathogenic bacteria in wastewaters may be destroyed by the process of chlorination.

The bacteriological safety of a wastewater is normally measured by the number of fecal coliform bacteria present. Coliform bacteria are not pathogenic but are an indication of the probability that pathogenic bacteria are present. Examples of pathogenic bacteria are *Salmonella, Shigella, Leptospira* and *Vibrio*. To this group of undesirable pathogens also can be added the enteric viruses and parasites, such as *Endamoeba hystolytica*.

Many of the monitoring stations now and in the future will be automated. The need for continuous monitoring for various parameters will determine how quickly this instrumentation becomes available. The nine regional offices of the Federal Water Quality Office (WQO), place nitrogen and phosphorous compounds, TOC, COD, and BOD at the top of their list.

TABLE 9

Automatic Monitoring Needs Identified by a WQO Water Quality Region Survey

Parameter	Number of Regions to Concur (of 9)
Nitrogen Compounds	100%
Phosphorus Compounds	100
BOD – COD – TOC Complex	100
Heavy Metals	66
Total Coliform	45
Fecal Coliform	45
Phenolics	45
Sulfate	33
Pesticides	33
Cyanide	33
Oil	33
Radioactivity	18
Alkalinity/Acidity	18
Chlorophyll	18
Lignin (PBI)	18

Source: Porterfield, N. W., Oceanology International, October 1970, p. 23.

The three (3) principal areas that should be discussed relative to monitoring are:

- Municipal Waste treatment plants
- Industrial waste facilities
- Stream surveys

The unit operations generally encountered in Municipal waste treatment (sewage treatment plant – STP) systems consist of:

OPERATION

- Screening
- Primary sedimentation
- Biological treatment
- Secondary sedimentation
- Tertiary treatment
- Ancillary systems
  - disinfection
  - sludge disposal
  - waste recirculation

CHART 2

Generalized Municipal Sewage Treatment Plant (S.T.P.) Flow Diagram

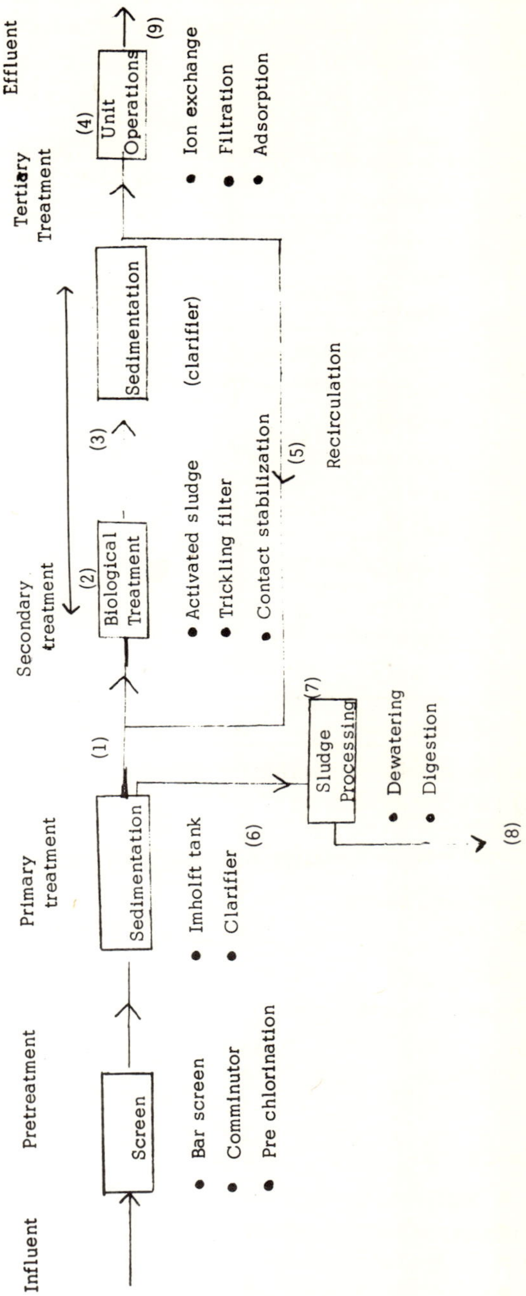

TABLE 10
Parameters for Monitoring at S.T.P.'s

PARAMETERS FOR MONITORING AT S.T.P.'S

Parameter	Primary/Secondary Treatment				Sludge
	Influent (1)	Biological Reactor (2)	Sludge Return (5)	Effluent (3)	Sludge Inlet (6)
Organic Loading (BOD)	o	o		o	
Phosphorus	o			o	
Nitrogen	o			o	
Toxicants	o				o
Dissolved Oxygen		o		o	
Residual Chlorine				o	
Gas Production/Composition					
Suspended Solids	o	o	o	o	o
Temperature	o	o		o	
pH	o	o		o	o
ORP		o			
Sludge Level			o		
Turbidity				o	
Color				o	
Conductivity				o	

NOTE: Numbers indicate location on flow diagram

Each of these steps may require monitoring for different parameters and at different frequencies. Table 10 lists suggested parameters for the monitoring of a STP.

MONITORED OPERATION OF S.T.P. [5]

Type of Test	Monitoring Method	Specific Monitoring Goals
Dissolved Oxygen	Electronic Polarographic	Used to control DO level in aeration tanks
Temperature	Thermistor Thermometer	Used to provide information for messaging the OU* value to STD conditions
Sludge Concentration	Photocell	Indication of operating of sludge return system
Oxygen Consumed	Short Term BOD	Designed to provide data for modifying the return sludge system operation to maintain the highest effluent quality.

*Oxygen Utilized

Table 10 (continued)

Digestion		Tertiary Treatment (4)							Final Effluent (9)
Digester (7)	Effluent (8)	Filtration	Coagulation	Air Stripping	Carbon Adsorption	Ion Exchange	Reverse Osmosis	Electrodialysis	
		o	o		o		o		o
			o			o			
				o		o			
									o
									o
o									
o	o	o			o				o
o									o
o				o					o
			o						
			o		o				
						o	o	o	

Industrial wastewater treatment is much more complicated than Domestic sewage. Domestic wastes are a quite uniform product, but industrial wastes scan the entire spectrum of inorganic and organic materials to toxic and radioactive materials. This, of course, complicates the treatment and monitoring systems used for operation discharge control.

Treatment in industrial waste practice centers around:

- physical methods
- chemical methods
- biological methods
- combination of treatment methods

Typical industrial waste processes were documented by W. W. Eckenfelder, in Chart 3. Typical processes used for industrial waste treatment in several industries are listed in Table 11. Pollutants that should be monitored at these same selected industrial operations are offered in Table 12.

Several volumes of text would be required for a full description of the innumerable details involved in the conduct of a successful stream pollution survey. Certain essential fundamentals, however, can be presented within the space constraints of this chapter.

There is a continuing need for pollution surveys to maintain surveillance of the waters of this country. This need arises from the increasing population as well as the continued industrial expansion, which oftentimes has not been accompanied by adequate waste treatment. Previously uninhabited areas are being saturated with industry and housing. As a result, more people are affected by pollution, actual or potential.

CHART 3

Processes Used for Industrial Waste Treatment

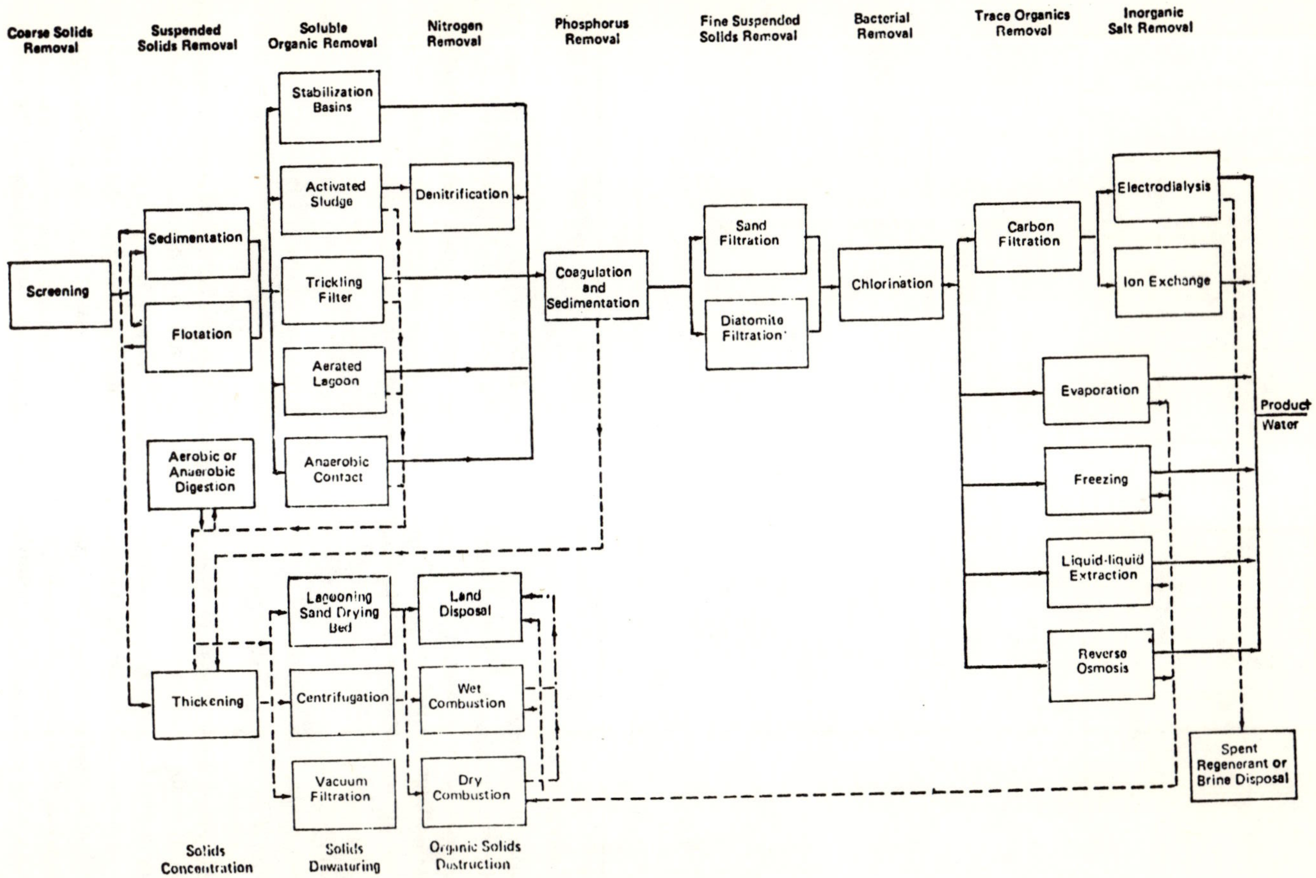

Source: Eckenfelder, W. W., Industrial Water Pollution Control, McGraw Hill, 1966

TABLE 11

Industrial Waste Treatment Technology

	PHYSICAL			CHEMICAL		BIOLOGICAL			
Industry	Sedimentation	Flotation	Coagulation	Oxidation/ Reduction	Neutralization	Activated Sludge	Trickling Filter	Lagoon	Spray Irrigation
Petroleum Refining	o	o	o		o	o	o	o	
Pulp and Paper	o	o	o			o	o	o	o
Fertilizer			o		o			o	
Textile Mills	o		o			o	o	o	
Metal Plating				o	o		o	o	
Vehicle Washing	o	o	o		o	o	o	o	o

TABLE 12

Effluent Parameters Monitored for Selected Industrial Waste Effluents

PARAMETER	INDUSTRY CATEGORY					
	Petroleum Refining	Pulp & Paper	Fertilizer	Textile Mills	Metal Plating	Vehicle Washing
Temperature	o	o	o	o	o	o
Suspended solids	o	o	o	o		o
pH	o	o	o	o	o	o
BOD	o	o				
COD	o		o	o	o	o
Iron					o	
Phenol	o					
Chlorides	o			o		
Free acid	o			o		o
Phosphates			o			
Oil	o					o
Sulfides/Sulfates	o	o				o
Fluorides			o			

Before we can proceed to abate pollution, it is necessary that we have evidence detailing the presence and degree of pollution that is in our streams.

OBJECTIVES OF A STREAM POLLUTION SURVEY

It is important to develop a specific objective and define it clearly before undertaking a stream pollution study. Some environmentalists appear to consider all stream pollution investigations as entirely new projects and fail to review available reports and operating records for background information that will assist in planning the surveys. Frequently, historical records provide information that is invaluable in the solution of a problem.

Because of the complexities of a river system, almost anything can happen, and frequently does; do not assume that a previous occurrence necessarily will be duplicated, but be alert to possible variations in the patterns of events.

It is our intent to set forth procedures to be followed so that more uniform and desirable reports will be obtained.

The major objective(s) of a stream survey generally is one (or more) of the following:

1. To determine the existing water quality of a stream (background studies).
2. To determine if the stream meets present or proposed water quality standards.
3. To determine the capacity of the stream to receive additional waste discharges.
4. To determine the environmental impact of a new industry on a stream.

Initial Preparation

The success or failure of a survey will depend to a large degree on the adequacy of the initial planning. Such planning should encompass the following:

1. Preliminary Reconaissance.—This should always be accomplished, to permit personnel in charge of the survey to become acquainted with the main features of the area in question. A limited number of samples should be collected to provide a preliminary indication of the range of concentrations to be expected. If bacteriological samples are to be collected, arrangements should be made at least seven days in advance with the laboratory.

Some engineers appear to think of a stream as a current of water flowing freely through space, and completely ignore the impact of the stream bed on the reactions of pollution in planning stream studies and in the interpretation of data. The concept may work reasonably well with large, deep streams in the absence of sludge deposits, but it falls apart completely where shallow, swift streams with rocky bottoms are involved. Observe and take into account all characteristics of the stream when planning a pollution survey.

2. Maps and Charts.—Necessary maps and/or charts must be secured. In some cases, a composite map must be constructed. Details such as coast lines, adjoining communities, sewer outfalls, bathing beaches and topographical features such as rivers, lakes, etc., including water depths, should be shown as required for clarity and understanding.

3. Selection of Sampling Stations.—Sampling stations must be selected to provide data which will, in turn, provide an adequate basis for determining the conditions of the area under study. In some cases, regular sampling stations may be included in the survey. Select stream sampling stations for accurate measurements of waste loads and of the orderly course of natural purification, and not primarily for saving time or convenience in sampling. When waste discharges, stream flow, or other stream conditions vary diurnally, around-the-clock sampling is highly desirable, if not imperative; in any event, sampling times should be varied throughout the daylight hours as much as is feasible.

4. Field Data.—Data to be obtained may vary greatly, depending on the scope and purpose of the survey. The areas that should be considered in gathering field data are:

a. Flora and fauna
b. Chemical quality
c. Bacteriological quality
d. Limnology
e. Benthic organisms

Natural phenomena should be included. These may include rainfall, tides, temperature and wind direction, and velocity.

One group of data often neglected is flows. It is important to remember that comprehensive evaluation of contaminant data frequently depends on the amount of contaminant discharged to a body of water, as well as on the concentration of that contaminant. Amounts cannot be determined unless the flow is known. Descriptions of flow-measuring devices should be included.

One of the most significant parameters, particularly as regards a report from a health agency, is the bacteriological pollution of surface and underground waters. Therefore, all such surveys should include sufficient bacteriological sampling, both as to sampling stations and number of samples to establish the degree and extent of such pollution.

It is important to stress the importance of determining sampling stations, sampling parameters, and number of samples to be collected before starting field surveys.

This is important because of:

1. financial constraints
2. purchasing of equipment
3. preparation of equipment
4. shipment of materials and equipment

In a recent case involving one of the largest streams in the United States, with untreated domestic wastewater and industrial wastes, a quick calculation was made using the familiar formula:

$$\text{mg/l} = \frac{\text{BOD (lb/day)}}{\text{Q (cfs)} \times 5.4}$$

showed that the known organic waste load would add only 0.2 to 0.3 mg/L of OBD to the stream. Another quick calculation with the less familiar formula:

$$\text{MPN/100 ml} = \frac{\text{Sewered Population}}{\text{Q (cfs)} \times (6.1 \times 10^{-5})}$$

indicated that the peak coliform density below town should be about 40,000 to 60,000 per 100 ml. The field crew was instructed to make a full bacteriological study, but not to determine BOD's unless a few preliminary sets of DO determinations revealed greater oxygen depletion than anticipated. If excessive DO depletion was found, additional laboratory facilities and personnel would be provided for a thorough BOD-DO study.

Carefully evaluate wastes that may cause water quality degradation and make all analyses necessary to show damage, but avoid the wasted energy and money consumed by analyses that do not contribute to proof of pollution. One or two sets of preliminary samples will indicate whether certain analyses will be productive in marginal cases.

5. Records.—It will be the responsibility of survey personnel to accurately and clearly record the results of all observations and tests to permit preparation of the corresponding portion of the report. Necessary field data sheets should be provided for survey personnel. These should be preserved until the final report is published and then forwarded to some type of inactive files.

6. Equipment.—After determining what is to be accomplished, check to see that necessary equipment and chemicals are on hand. This should include items such as boats, motors, necessary chemicals, sample bottles, sampling equipment and shipping cases, as well as flow-measuring equipment.

7. Duration.—Surveys should cover a minimum of five (5) sampling days unless the work is accomplished as a check on previous surveys. It is desirable to have the survey cover a period of seven (7) days if at all possible. As mentioned previously, in some cases a greater number of samples may be required. Get sufficient data to draw accurate conclusions and make reasonable recommendations.

Engineers rarely have an appreciation of the time required to process samples in the laboratory unless they have had laboratory experience themselves. The engineer who plans a survey without consulting his laboratory chief is almost certain to end up with a badly overloaded laboratory and a serious decrease in reliability of analytical results, to say nothing of a disgruntled laboratory crew. In planning a stream study, always obtain agreement with the laboratory chief on the number of samples that can be processed daily for the specified determinations by the personnel and facilities available.

Analysis

Tests are usually made in accordance with Standard Methods. As a rule tests should include pH, temperature, 5-Day 20°C BOD, dissolved oxygen, chlorides, total solids, suspended solids and settleable solids. It is desirable that total and, especially, suspended solids be further tested for fixed and volatile components. It is desirable that raw sewage and final effluent be tested bacteriologically. If chlorination is employed, the effluent should be tested for chlorine residual.

Tests ordinarily should be run on samples composited over a day's time. The sampling period should not be less than 8 hours unless it is definitely established there is little variation in volume and strength. Sixteen (16) hour composites are more desirable. Grab samples should be collected and analyzed if there is reason to suspect strong slugs of waste are being received.

When it is known that the stream contains industrial wastes, additional tests may be made to cover a wider range of pollutants. They may cover analyses such as nitrates, cyanides, chromium, ammonia and many others.

In some cases there will be a delay between collection and analysis. If dependable results are to be obtained, standard methods should be followed. In some instances a preservative must be used and the pH adjusted. This may require the collection of separate samples for different analyses. For example, the pH must be raised to a high level for cyanides and must be decreased to a fairly low level for chromium analysis.

OXYGEN DYNAMICS IN STREAMS

Dissolved oxygen is so economically, biologically, and technologically important that its significance is fairly universally recognized. Of the 50 states which prepared water quality standards in compliance with the Federal Water Quality Act of 1965 all employed dissolved oxygen (DO) concentration as a quality parameter.

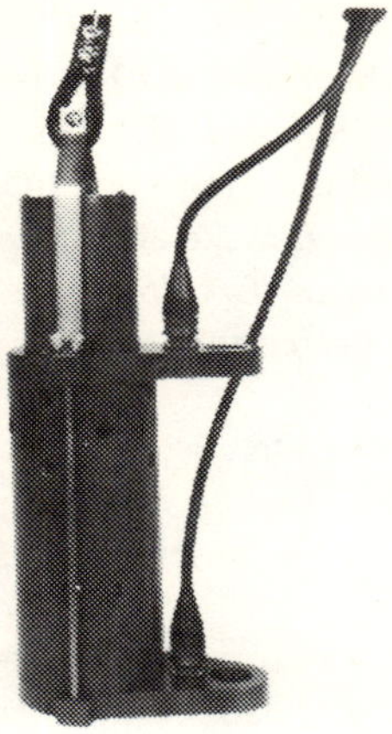

Figure 12. Dissolved oxygen sensor.
Source: Weston & Staell, Inc.

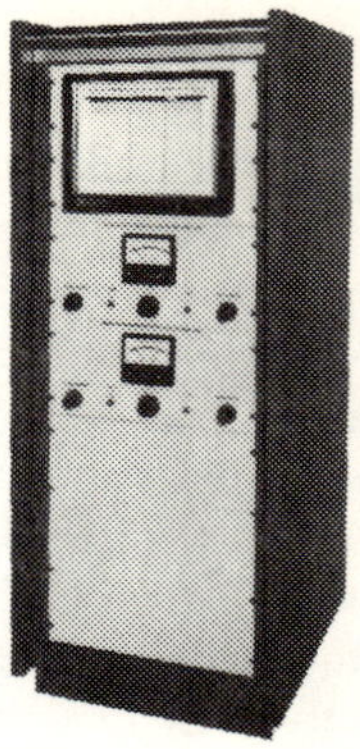

Figure 13. Recording instrumentation for dissolved oxygen measurements.
Source: Weston & Staell, Inc.

Effective water quality management is largely dependent upon control of the natural oxygen balance. Unless pollution decisions are based upon sound technical knowledge, achievement of desired water quality will be purely fortuitous.

Full comprehension of the natural oxygen balance is required to:

1. Predict the response of the oxygen resource to any stimulus (including pollution) before it occurs.
2. Assess the effectiveness of existing pollution control measures, and
3. Determine the degree of treatment required to maintain specific oxygen levels.

To efficiently control the oxygen balance one must: 1) account for all processes which influence DO resources, and 2) quantitatively estimate their effects.

The oxygen budget method requires measurement of all oxygen resources and liabilities under a given set of conditions. Should the same conditions reoccur at some future time the observed oxygen values would again prevail.

The budget method has been severely criticized because it tends to be laborious, evaluates the balance only under one set of environmental conditions, and presupposes that methods for independent measurement of the various assets and liabilities exist. Nevertheless, this method may be practical in uncomplicated situations.

The mechanism of oxygen exchange in natural waters could theoretically be described by a mathematical equation or model. A perfect mathematical model would account for all the myriad biological, chemical, and physical factors influencing the oxygen balance and could be used to predict oxygen levels under various conditions.

The mathematical model might be envisioned as a single equation containing numerous terms, each representing the influence of a chemical, physical, or biological variable. In practice, portions of this hypothetical equation have already been developed, but the elusive master equation itself is yet to be derived. At their present state of development, such mathematical models are not fully comprehensive but may be used provided the limitations of the analysis be known.

The values of coefficients and terms of the model may be estimated in the laboratory or derived from field observations.

The interactions which accomplish purification can best be understood by considering the various zones of pollution, with special attention to DO, BOD, and biological forms.

A. Clean Water Zone
   1. DO is near saturation
   2. BOD is very low (about 1.0 mg/l)
   3. There is a wide variety of biological forms, with relatively few of each form present.
B. Zone of Recent Pollution
   1. DO is slightly lower, and falling downstream
   2. BOD is at a maximum
   3. The organisms present are characteristic of polluted water. The variety is reduced, but there may be large numbers of a few forms.
C. Critical Zone
   1. DO is minimum, may go to zero
   2. BOD is lower
   3. Anaerobic organisms may be present if the DO has been completely exhausted.
D. Recovery Zone
   1. DO is rising, as reoxygenation has become the dominant process
   2. BOD is further reduced
   3. A larger variety of organisms is present, anaerobic forms are absent.
E. Clean Water Zone
   1. The stream has purified the waste lead imposed upon it
   2. Stream conditions are similar to A. above.

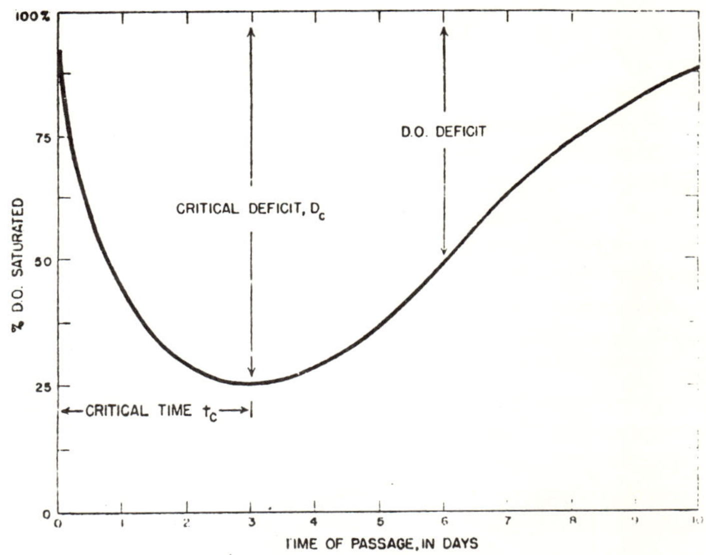

Figure 14. Oxygen sag curve.

VALUE OF NATURAL SELF–PURIFICATION

1. Utilization of the purification capacity of the stream permits installation of the most economical sewage treatment facilities.
2. The stream can be considered a part of the treatment plant, in that it accomplishes the final purification of the waste.
3. In order to avoid overtaxing the stream's purification capacity, an understanding of the factors involved and

an estimate of the reoxygenation potential is required.

4. A conservative estimate of the self-purification capacity will safeguard the stream water quality, yet permit economical sewage treatment.

Both the Water Quality Office (WQO) and the United States Geological Survey (USGS) have established guidelines for continuous monitoring stations. The typical areas for which specifications are described are:

- Retrieval of a sample (pumping)
- Precision of the instruments
- Range of parameters measured
- Physical arrangement
- Data recording capabilities

TABLE 13
Current Types of Instrumentation

Type Instrument		Range (ppm)	Approximate Cost ($)	Response Time	Accuracy	Comments
Electrometric Instruments	Dissolved Oxygen Analyzers	0–100	400–1000	< 1 min	–	–
	pH/ORP Analyzers	–	150–1500	–	0.02–.5	–
	Specific Ion Analyzers	1 ppb-10,000 ppm	900–3500	–	–	Up to 16 ions
	Conductivity Analyzers	<100->10,000	100–8000	–	1–2 o/o (full scale)	Range in μmho
	Res. Cl_2	0–1 0–20	600–3500	5–15 sec	–	–
Suspended Solids Analyzers	–Low/Medium Turbidity	0–.2 0–5000	100–1800	–	–	–
	–Low Level	0–1 0–50,000	200–6000	–	–	–
	–Sludges	0–10 o/o	3000–5000	–	–	–
Automated Wet Chemical Analyzers		–	1600–30,000	–	–	–
Dissolved Carbon/Oxygen Demand Analyzers		0–10 0–5000	5000–10,000	–	–	TOC, BOD, TC, COD
Optical Analyzers		–	700–5000	–	–	For Dissolved Species
Particle Counters		0.1–2500 microns	6000–20,000	–	–	–

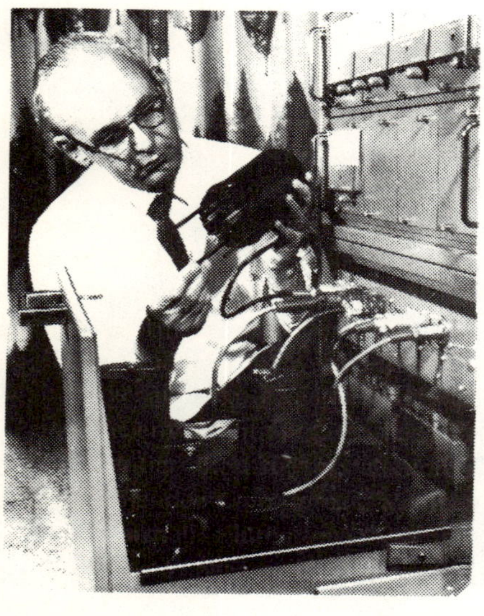

Figure 15. Monitoring sensor being checked.
Courtesy: Honeywell Industrial Division
1100 Virginia Drive
Fort Washington, Pennsylvania 19034

Figure 16. Water Pollution Patrol Boats in St. Louis.
Courtesy: Honeywell Industrial Division
1100 Virginia Drive
Fort Washington, Pennsylvania 19034

Figure 17. Water quality monitoring system onsite — sensing & automatic analog recording.
Courtesy: Honeywell Industrial Division
1100 Virginia Drive
Fort Washington, Pennsylvania 19034

One of the parameters for stream pollution work that can now be monitored is biomass. The instrumentation was designed for the measurement of the bioluminescent flash caused by the reaction of Adenosine Triphosphate (ATP) and luciferin in the presence of the firefly luciferase enzyme. Biomass determinations in many different types of media can be rapidly and accurately performed for a wide range of applications.

The overall reaction is given by:

$$ATP + O_2 \xrightarrow[\text{luciferase}]{M++} \text{luciferin (oxidized)} + pp_i + AMP + H_2O + h\nu$$

When a prepared sample containing ATP is introduced into the enzyme preparation, there is an immediate burst of light in the wavelength range of 560 to 580 nm. The intensity of the light decays in an exponential fashion.

Rather than measure the flash peak height with its inherent reproducibility problems, the JRB unit delays for 15 seconds then measures and integrates the area under the light curve for the next 60 seconds. The counts per minute (CPM) can be compared with a standard calibration curve to determine the absolute quantity of ATP present in the sample. This innovative approach allows for mixing of reactants outside of the light chamber and the use of the cheaper nonrefined enzyme. Disposable scintillation vials can be used as reaction vessels.

The object of the preparation procedure is to:

- Extract the ATP from the organisms by killing the cells,
- Deactivate the ATP-ase which would degrade the released ATP to Adenosine Di-phosphate and Adenosine Mono-phosphate,
- Separate the ATP solution from cellular and extra-cellular debris.

The most common procedure is to lyse the isolated cells by boiling in a tris buffer solution (pH 7.7). Direct aliquots can be introduced into the boiling tris, or, if the level or organisms is low, one can concentrate the cells by filtering. Other standard lysing techniques that have been used are: perchloric acid, butanol, trichloroacetic acid, arsenate buffer and DMSO. Sample preparation and lysing usually takes from 4 to 10 minutes.

Figure 19 entitled "ATP Photometer (Biomass Determinations) is furnished courtesy of J.R.B., Inc., LaJolla, California.

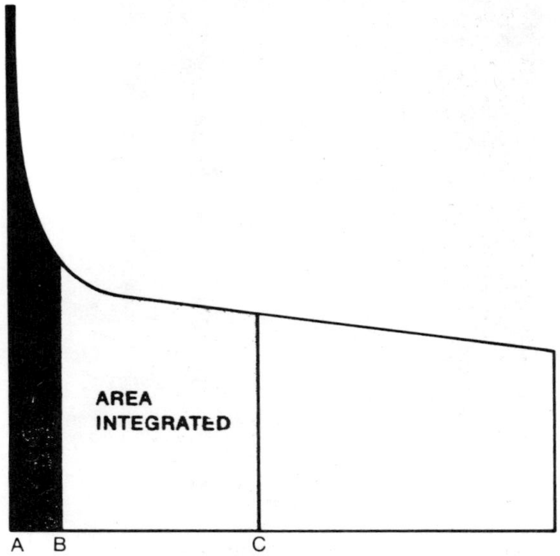

Figure 18. Light emission response when a sample containing ATP is injected into the enzyme preparation: pt. A, time of injection of sample; pt. B, end of 15 second delay period; pt. C, end of one minute light-integration period.

Courtesy: J. R. B., Inc.

Figure 19. ATP Photometer (Biomass determinations).

Courtesy: J. R. B., Inc.
LaJolla, CA

The biomass determination advantageously utilizes the following characteristics of ATP:

- ATP is found in all living cells,
- ATP is not found associated with nonliving particulate matter,
- ATP is found in a relatively constant concentration in all living cells,
- ATP can be quantitatively assayed in extremely low concentrations by using the chemiluminescent reaction.

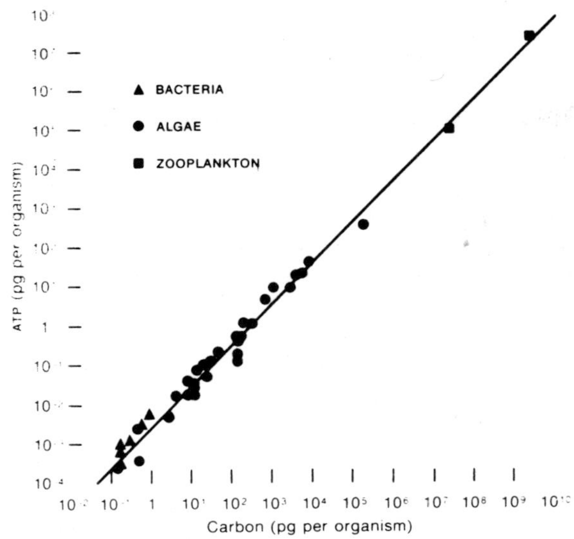

Figure 20. Cellular contents of ATP as a function of total organic carbon for various unicellular and multicellular organisms.

Courtesy: J. R. B., Inc.

Total carbon analysis by standard combustion techniques in conjunction with the ATP measurement can thus establish live-to-dead cell carbon ratios.

The ATP Photometer has many potential applications in environmental quality and environmental control. It gives rapid, accurate measurements of the quantities of algae, zooplankton, and bacteria in solution.

An index of thermal pollution is possible through quantification of the effect of temperature increase on algae or plankton populations.

The photometer can be employed as a process control parameter in sewage treatment and pollution control to monitor the concentrations of break-down organisms in activated sludge.

Data obtained through a well planned and executed monitoring program will provide valuable information to those individuals responsible for the proper operation of plant processes and environmental control. The data obtained from a monitoring system can be employed in the evaluation and possible alteration of in-plant and wastewater treatment processes and may influence the commitment of large capital expenditures. Thus, the parameters monitored and the significant results obtained from the monitoring program must be critically evaluated prior to alterations of the processes involved. Because of the significance which may be placed on the results of the monitoring program, it should be the desire of plant management to distinguish between the data results which are influenced by fluctuations which are not representative of the discharges from the processes involved. Placing significance on monitoring results which are not representative may result in unnecessary expenditures or in a false sense of security.

The variability of the parameters measured in the monitoring system may result from various phenomena, some of which are listed below.

1. In-plant spills or poor housekeeping
2. Temporary modification of in-plant processes
3. Chemical reactions resulting from various combinations of waste discharges
4. Improper maintenance and/or operation of monitoring equipment
5. Errors in calculations or analysis or measured parameters.

Those deviations which are not representative of the industrial waste treatment processes employed must be identified prior to the evaluation of the monitoring results. They may indicate the need for improvement in the maintenance and operation of the industrial processes, waste treatment processes, or the monitoring system itself.

The variability of the parameters may be random, resulting from the random effects on the process and its measurement, or cyclic, resulting from periodic phenomena affecting the process (daily records, weekly periods, etc.).

The purpose of a wastewater survey is to obtain sufficient data about the wastewater characteristics of a plant so that a monitoring program can be established. In obtaining data, the values for various wastewaters will be observed to vary or fluctuate with time, location, sample, analysis, personnel, type of preservation and perhaps a few more less evident factors. Time is usually considered the most significant factor causing fluctuation in data. Another important factor is location. Samples taken at the same time but at different locations, even in the same sewer, often present different results. The preliminary investigation should minimize these sources of errors. The results of the analysis of a sample, by the same or different technicians and using the same laboratory techniques, often fluctuate widely. Even very accurate laboratory analysis cannot prevent a relatively wide range in determined values of parameters, such as BOD.

Statistics aid in the development of general laws resulting from numerous individual determinations which, by themselves, may be meaningless. The resulting relationships are part of the fundamental function of statistics which expresses the data obtained from an investigative process in a condensed and meaningful form. Thus, the average or mean is often used as a single value to represent a group of data. The variability of the group of observations is expressed by the value of the standard deviation and trends in concentrations measured during the monitoring process are expressed in the form of regression coefficients.

In general, the concern is with the treatment of the collected data. The accuracy or usefulness of these data is greatly enhanced if a full understanding was involved in generating the facts. The balance between use of statistical methods and evaluation based upon physical understanding is extremely important. The use and value of statistics decreases as physical understanding increases. Specifically, the difficulty lies in separating chance effects from valid occurrences. With the knowledge of basis probability theory and the use of statistical techniques, such as Least Squares Curve Fitting, Analysis of Variance, Regressive and Correlation Analysis, Chi-Squared Goodness of Fit, and others, it is possible to construct mathematical models and curves of almost any level of precision desired. Such techniques help to evaluate information having wide variations, so that an estimate of the best value of the parameter being measured can be assigned; and also to assess the precision of that estimate. Statistical procedures may also help in identifying errors and mistakes and are helpful in comparing sampling methods and procedures and in evaluating waste loadings from different process schemes.

FUTURE TRENDS

Many estimate the water monitoring equipment market as continuing to expand well past 1985.

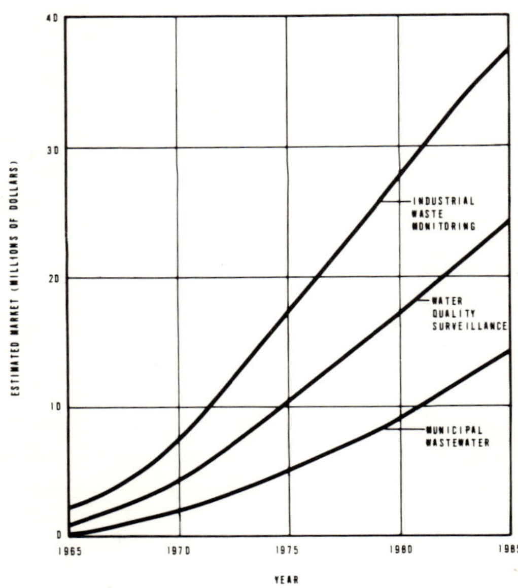

Figure 21. Future trends in monitoring equipment (estimated).

We still have a substantial need for highly sensitive measuring and continuous monitoring techniques for:

- Trace metals
- Pesticides
- Phenol(s)
- Nutrients
- Micro-organisms

Indicative of the current status of monitoring equipment is the instrument needs for water quality monitoring.

Table 14 lists those parameters and variables which may be considered necessary for: qualification of water source; selection of treatment method; regulatory health requirements; quality goals; and the control (or potential use in the control) of the water-treatment process.

TABLE 14

Status of Automating Water Quality Analyses [7]

Variable or Parameter	*Instrument Available*	*Instrument Near Term*	*Manual Now*
Alkalinity	x		
Aluminium		x	
Ammonia	x		
Arsenic		x	
Bacteria			
Barium		x	
Beta-radio-gross	x		
BOD			x
Cadmium		x	
Calcium		x	
Carbon alcohol extract			x
Carbon chloroform extract			x
Carbon dioxide, free		x	
Carbon, total organic		x	
COD		x	
Chloride	x		
Chlorine, free	x		
Chlorine, total	x		
Chlorine demand			x
Chromium, hexavalent	x		
Coliforms			x
Colloidal titration		x	
Colour		x	
Conductivity	x		
Copper	x		
Cyanide		x	
Fluoride	x		
Hardness	x		
Lead		x	
Magnesium		x	
Manganese		x	
Mercury		x	
Nitrogen, nitrate	x		
Nitrogen, nitrite	x		
Nitrogen, organic	x		
Odour			x
Oil		x	
Oxygen reduction potential	x		
Oxygen, dissolved	x		
Oxygen, per cent saturation	x		
*p*H	x		
Phenol		x	
Phosphates, meta	x		
Phosphates, ortho	x		
Phosphates, total	x		
Potassium		x	
Residue, filtrable			x
Residue, total			x
Selenium		x	
Silica	x		
Silver		x	
Sodium		x	
Streaming potential		x	
Sulphate	x		
Surfactants	x		
Taste			x
Temperature	x		
Turbidity	x		
Zeta potential		x	
Zinc		x	

BILBIOGRAPHY

1. Vernon T. Stack, Monitoring in ***Water Quality Control & Management***, presented at Technicon International Symposium, Sydney, Australia, February, 1971.
2. F. W. Kittrell, "Conducting Water Pollution Surveys," Stream Sanitation Studies, Water Pollution Control, Robert A. Taft Sanitary Engineering Center, Cincinnati, Ohio.
3. EPA Ref.
4. F. L. Mascitti, "Modern Instrumentation and the Challenges of Water Problems," International Conference on Water for Peace, May 23–31, 1967, Washington, D.C.
5. R. S. Ingols and R. H. Morriss, "Control Monitoring for the Activated Sludge Process,",***Deeds & Data***, supplement to ***Highlights***, WPCF, August 1973.
6. R. S. Ingols, "Short Term BOD," Water & Sewer Works, 115, 258 (1968).
7. J. V. Radziul and I. H. Suffet, "Automation of Water Analysis and Its Use to Control Water Purification Processes," International Water Supply Congress, August, 1972.